FACT FINDERS
KNOW THE DIFFERENCE!

ASTEROID OR COMET?

EZRA E. KNOPP

PowerKiDS press.

Published in 2026 by The Rosen Publishing Group, Inc.
2544 Clinton Street, Buffalo, NY 14224

First Edition

Editor: Greg Roza
Book Design: Michael Flynn

Photo Credits: Cover (asteroids) Vadim Sadovski/Shutterstock.com; cover (comet) Lauritta/Shutterstock.com; p. 5 AstroStar/Shutterstock.com; p. 6 YummyBuum/Shutterstock.com; p. 7 Artsiom P/Shutterstock.com; p. 8 https://commons.wikimedia.org/wiki/File:Costanzo_Angelini,_L%27astronomo_Piazzi_1825_ca.jpg; p. 9 https://commons.wikimedia.org/wiki/File:Ceres_01.jpg; p. 10 https://commons.wikimedia.org/wiki/File:Bayeux_Tapestry_32-33_comet_Halley_Harold.jpg; p. 11 Brian Donovan/Shutterstock.com; p. 12 PintoArt/Shutterstock.com; p. 13 Shutterstock AI Generator/Shutterstock.com; p. 14 Nazarii_Neshcherenskyi/Shutterstock.com; p. 15 Blueee77/Shutterstock.com; p. 16 joshimerbin/Shutterstock.com; p. 17 https://commons.wikimedia.org/wiki/File:Rosetta_and_Philae_at_comet_(11206660686).jpg; p. 18 Jan Garbers/Shutterstock.com; p. 19 Alfonso Blanco/Shutterstock.com; p. 21 AstroStar/Shutterstock.com.

Cataloging-in-Publication Data

Names: Knopp, Ezra E.
Title: Asteroid or comet? / Ezra E. Knopp.
Description: Buffalo, New York : PowerKids Press, 2026. | Series: Fact finders: know the difference! | Includes glossary and index.
Identifiers: ISBN 9781499452662 (pbk.) | ISBN 9781499452679 (library bound) | ISBN 9781499452686 (ebook)
Subjects: LCSH: Asteroids--Juvenile literature. | Comets--Juvenile literature.
Classification: LCC QB651.K56 2026 | DDC 523.44--dc23

Manufactured in the United States of America

CPSIA Compliance Information: Batch #CSPK26. For Further Information contact Rosen Publishing at 1-800-237-9932.

CONTENTS

PEERING INTO SPACE

There's so much to learn about space! For thousands of years, scientists called astronomers have looked at the night sky in search of answers to **cosmic** questions. New inventions—such as telescopes and spacecraft—have led to many answers about space. But there's so much more to learn!

Have you ever wanted to know more about space and astronomy? Learning about asteroids and comets is a good place to start! Do you know how these two things are the same and different? Let's **explore** space and find out!

ASTRONOMY IS THE STUDY OF OBJECTS IN SPACE, SUCH AS PLANETS AND MOONS. THIS ASTRONOMER IS USING A TELESCOPE TO MAKE A DISTANT, OR FARAWAY, PLANET SEEM CLOSER AND LARGER SO HE CAN STUDY IT.

THE BIRTH OF THE SOLAR SYSTEM

Our solar system formed 4.6 billion years ago! The solar system is made up of the sun, the planets that **orbit** it, and the moons that orbit the planets. Which formed when the solar system first formed, asteroids or comets?

The answer is both! Asteroids and comets are both made up of the same matter that makes up the planets. The sun formed from a giant cloud of dust. The planets formed soon after. However, some matter was left over. That matter formed asteroids and comets.

DISCOVERY!

On January 1, 1801, the Italian scientist Giuseppe Piazza observed a new light in the night sky. Other scientists studied the light and soon figured out what it was. Did they see an asteroid or a comet?

GIUSEPPE PIAZZA

SCIENTISTS HAVE FOUND MANY OTHER ASTEROIDS SINCE 1801. IN 2006, SCIENTISTS DECIDED THAT CERES WAS BIG ENOUGH TO BE CALLED A **DWARF PLANET** INSTEAD OF AN ASTEROID.

Piazza was the first person to observe an asteroid! At first, he thought it was a distant star. But other scientists figured out that the asteroid was orbiting the sun at an even distance. Ceres, as the asteroid was soon named, was the biggest asteroid in the solar system!

BEYOND THE SOLAR SYSTEM

Scientists have recorded lights in the night sky for hundreds of years. Some observed that one passing light had visited the sun many times, returning about every 75 years. Were they witnessing an asteroid or a comet?

AT THE TIME OF THIS WRITING, HALLEY'S COMET JUST BEGAN ITS LONG JOURNEY BACK TOWARD THE SUN. WHEN NEAR, PEOPLE ON EARTH CAN SEE THE COMET IN THE SKY WITHOUT A TELESCOPE!

It was Halley's Comet! This comet, like all comets, makes an **elliptical** path that takes it past the edge of the solar system! Asteroids, on the other hand, are mainly found in orbit between Mars and Jupiter in an area called the asteroid belt.

TALL TAILS

When scientists spot a new light in the **universe**, they study how it acts to find out what it is. There's a key clue that allows them to find out if the light is a comet or an asteroid. One of them has a tail!

Asteroids are mostly rock, but comets are made of frozen rock, ice, and gas. As a comet nears the sun, some of this frozen matter melts. The comet leaves these things behind it, forming what looks like a tail.

SPACE TOTALS

Scientists have discovered nearly 4,000 comets! They believe there could be many more far outside the solar system they haven't spotted yet. Which do you think scientists have counted more of: asteroids or comets?

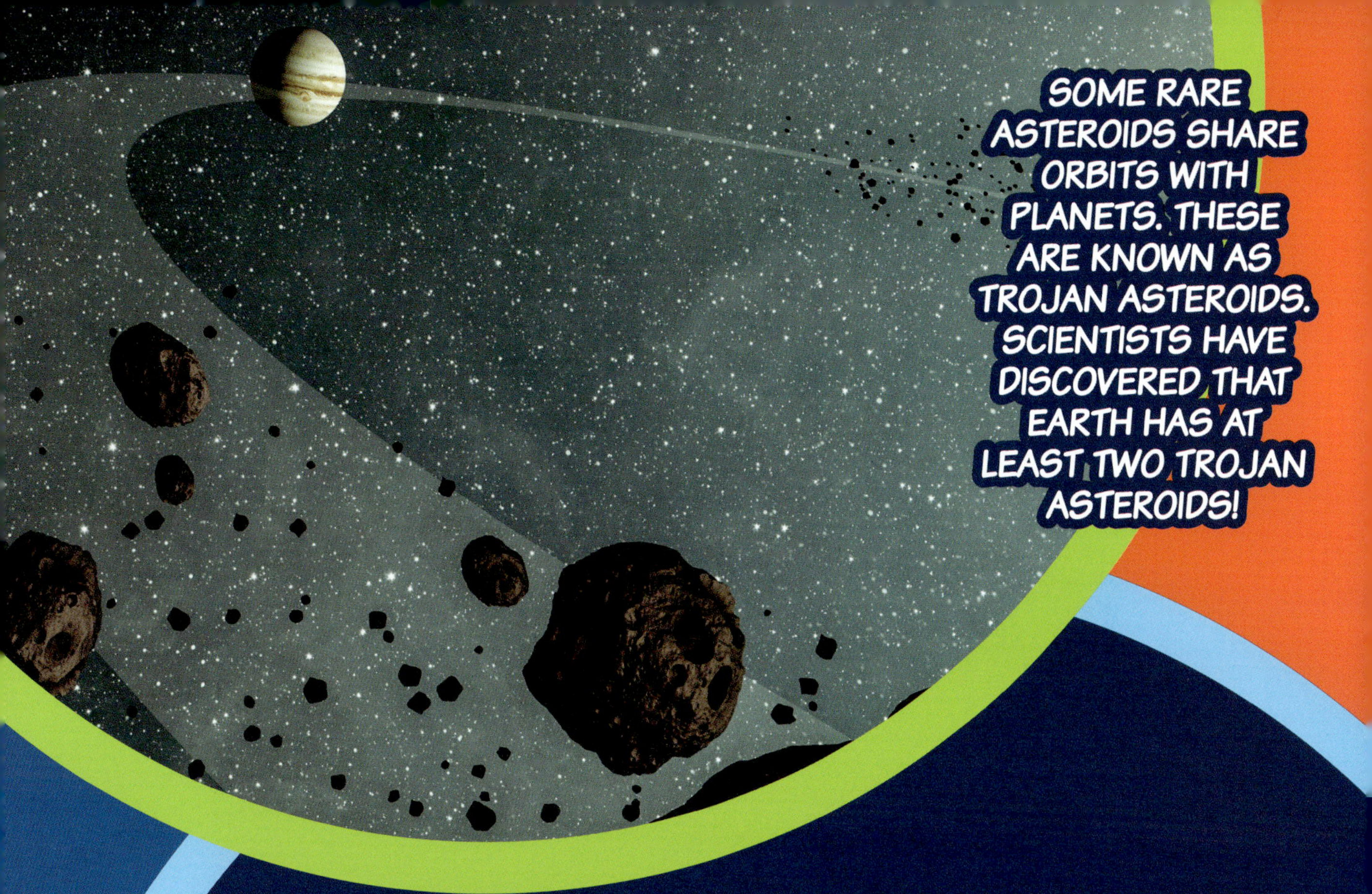

Most asteroids orbit the sun between Mars and Jupiter. There are so many asteroids in the asteroid belt that scientists can only guess how many are there—between 1.1 and 1.9 million asteroids! The total **mass** of the asteroid belt is less than that of our moon.

PROBING THE SOLAR SYSTEM

Scientists are always studying and exploring space. They've even sent out space probes to learn more! Space probes are unmanned craft that have tools and **instruments** that send information back to scientists on Earth. Can probes be used to study asteroids and comets?

OSIRIS-REx

Yes! Scientific teams on Earth have landed probes on a comet and an asteroid! In 2014, the Rosetta probe sent a tinier probe down to a comet's surface. In 2020, the OSIRIS-REx probe landed on and collected **samples** from the asteroid Bennu.

SHOOTING STARS

A shooting star is a flash of light in the night sky. They move quickly and don't last for long. Are shooting stars asteroids or comets? Trick question! Shooting stars aren't asteroids or comets. They're meteors.

Meteoroids are small pieces of rock that used to be part of an asteroid or comet but broke apart. When these pieces enter Earth's **atmosphere** they are called meteors. They quickly burn up while falling toward the ground. Many meteors falling at one time is called a meteor shower.

KEEP EXPLORING!

We've used our observation skills to make some important discoveries about asteroids and comets. We even learned some things about the solar system, meteors, and space probes. Of course, we needed to use **technology** to find those answers.

Do you have any other questions about asteroids and comets? What about space in general? How do you find the answers to those questions? You've made a great start by reading this book! But don't stop there: Keep exploring the universe!

TELESCOPES ARE AN IMPORTANT TOOL IN ASTRONOMY. HOW DO THEY MAKE FARAWAY OBJECTS SEEM CLOSER AND LARGER? SOUNDS LIKE ANOTHER FUN SCIENCE TOPIC FOR YOU TO EXPLORE!

GLOSSARY

atmosphere: The mass of air that surrounds Earth or another planet.

cosmic: Having to do with the cosmos, or outer space.

dwarf planet: A body in space that orbits the sun and is round like a ball but isn't large enough to disturb other objects from their orbit.

elliptical: Describing the orbit of a body that moves in an oval path.

explore: To search an area in a careful way to learn more about it.

instrument: A tool or computer made to do a very specific job.

mass: The amount of matter than makes up something.

orbit: The path one object in space takes around another.

sample: A small amount of something collected from a larger mass, used for scientific study.

technology: A method that uses science to solve problems and the tools used to solve those problems.

universe: Everything that exists.

FOR MORE INFORMATION

BOOKS

Betts, Bruce. *Asteroids and Comets: Orbiting Space Rocks*. Minneapolis, MN: Lerner Publications, 2025.

Lombardo, Jennifer. *Comets and Asteroids*. Buffalo, NY: Cavendish Square, 2024.

Nelson, Louise. *The Solar System*. Minneapolis, MN: Jump!, Inc., 2024.

WEBSITES

NASA: What Is a Comet?
spaceplace.nasa.gov/comets/en
Read more about comets at NASA's Space Place!

NASA: What Is an Asteroid?
spaceplace.nasa.gov/asteroid/en
NASA's Space Place is also a great place to learn about asteroids.

INDEX